YOUR KNOWLEDGE HAS VALUE

Laura Imperatori

Mechanical Resonance. Free and forced SHM of a torsional pendulum

GRIN Verlag

Bibliografische Information der Deutschen Nationalbibliothek:

Die Deutsche Bibliothek verzeichnet diese Publikation in der Deutschen National-
bibliografie; detaillierte bibliografische Daten sind im Internet über http://dnb.d-
nb.de/ abrufbar.

Dieses Werk sowie alle darin enthaltenen einzelnen Beiträge und Abbildungen
sind urheberrechtlich geschützt. Jede Verwertung, die nicht ausdrücklich vom
Urheberrechtsschutz zugelassen ist, bedarf der vorherigen Zustimmung des Verla-
ges. Das gilt insbesondere für Vervielfältigungen, Bearbeitungen, Übersetzungen,
Mikroverfilmungen, Auswertungen durch Datenbanken und für die Einspeicherung
und Verarbeitung in elektronische Systeme. Alle Rechte, auch die des auszugsweisen
Nachdrucks, der fotomechanischen Wiedergabe (einschließlich Mikrokopie) sowie
der Auswertung durch Datenbanken oder ähnliche Einrichtungen, vorbehalten.

Imprint:

Copyright © 2012 GRIN Verlag GmbH
Druck und Bindung: Books on Demand GmbH, Norderstedt Germany
ISBN: 978-3-656-58599-2

This book at GRIN:

http://www.grin.com/en/e-book/268472/mechanical-resonance-free-and-forced-
shm-of-a-torsional-pendulum

Mechanical Resonance

- Free and forced SHM of a torsional pendulum

Laura Imperatori, Murray Edwards College, lsi22

Experiment performed Friday, 27/01/2012

(Practical partner x, Trinity College)

Abstract

In order to test SHM, the behaviour changes of a torsion pendulum due to different damping factors as well as its changes due to applying an external exciter were observed and compared with the theoretical expectations. The quality factor of the same damping state (with a brake current of 6A in the eddy brakes) was calculated using two different approaches and the resulting values were found to be within σ of each one, as Q_1 =6.0±1.3 and Q_2 =7.5±0.7. The first approach was based on measuring the maximum displacement for each successive oscillation and deducing the slope from the plot of the natural logarithm of the amplitude against the number of oscillations. Differently, the second estimate of Q was obtained under forced oscillation conditions by taking the ratio of the experimentally determined resonance amplitude and the amplitude of natural oscillation.

I. Introduction

The aim of the experiment is to measure free, damped and forced SHM in a torsion pendulum. The investigation of damped oscillations led us to the determination of the quality factor Q, while the examination of forced oscillations resulted in observations of resonance phenomena and thus the determination of the resonance frequency. By combining the damped and forced oscillations, it was possible to investigate how the degree of damping affects the amplitude of the response to a sinusoidally varying spring force.

In principle, the quality factor Q can be deduced from any damped SHM oscillating system by measuring the natural frequency ω_0 and the decay constant γ as Q equals $\frac{\omega_0}{2\gamma}$. Resonance can also be observed in many different systems, involving simple pendulums and string pendulums. As these measurements are very inaccurate and imprecise, we used a special torsional pendulum developed by German physicist Robert Wichard Pohl (see VI.2). This pendulum is

1

damped with a manually variable eddy current brake that is widely used for didactic purposes demonstrating mechanical resonance to undergraduate Physics students.

Current research also involves the use of torsional pendulums. For example, the Eöt-Wash Group at the Center for Experimental Nuclear Physics and Astrophysics of the University of Washington tests the equivalence principle of Einstein's General Theory of Relativity with torsional balances. (See VI.3)

II.   Theoretical Background

Simple harmonic oscillation of a system can be caused by any restoring force (or torque) which is directly proportional to linear (or angular) displacement. In the case of a torsion pendulum like the one investigated in this experiment, the restoring torque is supplied by a spring. Along with Newton's Second Law, this suggests a second order differential which describes the oscillating motion of an object of moment of inertia I suspended on a torsion wire of torsion constant τ:

$$I\ddot{\theta} = -\tau\theta$$

where θ is the angular displacement from equilibrium.

The solution is conventionally written: $\theta = \theta_0 \cos(\omega_0 t)$ with the natural frequency of oscillation $\omega_0 = \sqrt{\frac{\tau}{I}}$ under the assumption that the system has been released from rest at an angle θ_0 at t=0.

If there is an opposing damping force proportional to the angular velocity $\dot{\theta}$, the SHM equation in (1) has to be modified to be

$$I\ddot{\theta} = -\tau\theta - b\dot{\theta}$$

where b is a constant.

This can be generalised to be

$$\ddot{\theta} + 2\gamma\dot{\theta} + \omega_0^2\theta = 0 \quad (1)$$

with the decay constant $\gamma = \frac{b}{2I}$ as a measure of damping and $\omega_0 = \sqrt{\frac{\tau}{I}}$ as natural frequency. The motion of the pendulum is now decaying according to:

$\theta = Re\{Be^{pt}\} = \theta_0 e^{-\gamma t} \cos(\omega_1 t)$, where $\omega_1 = \sqrt{\omega_0^2 - \gamma^2}$.

2

Depending on the parameters B and p, which vary with the magnitude of the damping coefficient b compared with I and τ, there are three different resulting cases called underdamped, overdamped and critically damped.

The plot of amplitude against time of these three different cases is shown in figures F1, F2 and F3:

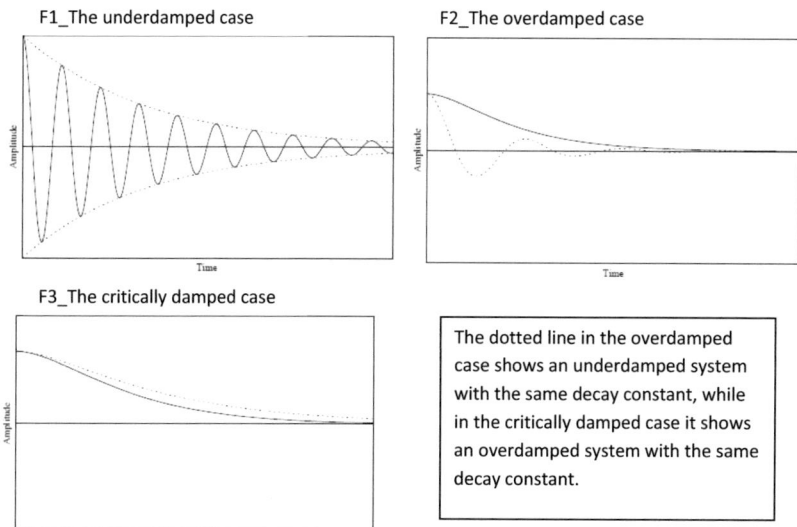

F1_The underdamped case

F2_The overdamped case

F3_The critically damped case

The dotted line in the overdamped case shows an underdamped system with the same decay constant, while in the critically damped case it shows an overdamped system with the same decay constant.

As visible in the above graphs, the only case that oscillates is the underdamped case. Its free response is of the form $\theta = \theta_0 e^{-\gamma t} \cos(\omega_1 t)$, where $\omega_1 = \sqrt{\omega_0^2 - \gamma^2}$ with the decay constant γ. The slower the decay of the oscillations in relation to the period, the better is an oscillator. Hence, for a good oscillator the quality factor as measure of this has to be significantly greater than 1: $Q = \frac{\omega_0}{2\gamma} \gg 1$.

In the case of additionally applying a sinusoidal driving couple $A \cos(\omega t)$, the overall equation becomes:

$$I\ddot{\theta} = -\tau\theta - b\dot{\theta} + A \cos(\omega t) \text{ or simplified } \ddot{\theta} + 2\gamma\dot{\theta} + \omega_0^2\theta = \frac{A}{I}\cos(\omega t). \text{ (2)}$$

The general solution to the above equation (2) is the superposition of a transient (or free) and steady-state (or forced) response: the sum of the solution of (1) and a particular solution to (2).

The particular solution can be found by solving equation (2) of the form

3

$$\theta(\omega) = X(\omega)\cos(\omega t - \varphi(\omega)).$$

Direct substitution gives the amplitude as $X(\omega) = \dfrac{A/_I}{\sqrt{(\omega_0^2-\omega^2)^2+(2\gamma\omega)^2}}$ and $\tan[\varphi(\omega)] = \dfrac{\gamma\omega}{(\omega_0^2-\omega^2)}$,

where $\varphi(\omega)$ is the phase difference between the driving couple and the oscillatory response of the pendulum.

The behaviour of $X(\omega)$ can be summarised as follows:

1. The amplitude in the limit as $\omega \to 0$ and as $\omega \to \infty$ can determined to be:
   a. $\lim_{\omega \to 0} X(\omega) = \dfrac{A/I}{\omega_0^2}$,
   b. $\lim_{\omega \to \infty} X(\omega) = 0$.
2. The amplitude has a maximum at $\omega_{max}^2 = \omega_0^2 - 2\gamma^2$, where ω_{max} is called the amplitude resonant frequency. Its limiting value in case of very low damping is ω_0.
3. At ω_{max} the amplitude is given by $X(\omega_{max}) = \dfrac{A/_I}{2\gamma\sqrt{(\omega_0^2-\gamma^2)}}$.
4. In the limit of very low damping (where $\gamma \ll \omega_0$), γ^2 is insignificantly small. Hence, the amplitude at resonance $X(\omega_{max})$ can be regarded as $X(\omega_{max}) \sim \dfrac{A/_I}{2\gamma\omega_0}$ which equals Q times the amplitude when $\omega \to 0$:

$$Q \times X(\omega \to 0) = X(\omega_{max}) \sim \dfrac{A/_I}{2\gamma\omega_0}.$$

III. Methods and Results

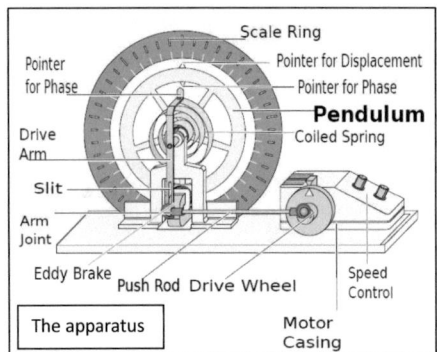

Pointer for Phase
Scale Ring
Pointer for Displacement
Pointer for Phase
Pendulum
Drive Arm
Coiled Spring
Slit
Arm Joint
Eddy Brake
Push Rod Drive Wheel
Speed Control

The apparatus
Motor Casing

The torsion pendulum used in this experiment is a bronze disc which undergoes rotational motion. The restoring force provided by the coiled spring is linearly proportional to the angular displacement of the pendulum. This can be measured with the scale of arbitrary units on the

4

outer annulus of the "Pohl wheel" (see VI.2). Oscillations of the pendulum can be driven sinusoidally by means of a push rod and drive arm connected at one end to the pendulum through the coiled spring and the other to a rotating drive wheel. The speed of the drive wheel and therefore the period of the forced oscillation can be controlled by the voltage supplied, as measured by a voltmeter. Damping beyond the unavoidable imperfections in the system is provided by an adjustable eddy brake which exerts a force proportional the angular velocity of the pendulum.

1.) Transient response

i.   "Zero Damping" Behaviour

First of all, the behavior of the system was checked when there was no current passing through the eddy brakes. Due to internal friction and the air resistance there was still a degree of damping in the oscillatory motion of the pendulum. The amplitude slowly decayed from its furthest elongation at 10 on the scale ring to its equilibrium position.

ii.   Natural frequency measurement

The natural frequency of the pendulum ω_0 was estimated by four consecutive measurements of the time needed for ten full oscillations of the pendulum when displaced and released. Hence, the period of the pendulum could be determined to be $T = 1.916 \pm 0.001 s$, which corresponds to a natural frequency $\omega_0 \sim \omega_1 = \frac{2\pi}{T} = 3.280 \pm 0.014 \frac{rad}{s}$ considering the unavoidable damping as small enough to be negligible.

iii.   Analysis of the free oscillation of the pendulum for two degrees of damping

For each successive displacement n ($1 \leq n \leq 15$) the maximum displacement a_n was recorded in two different cases − a) under no influence of the eddy brakes and b) with a braking current $I_b = 0.6A$. As the exponentially decaying envelope of the free response can be described with $a_n = a_0 e^{-n\gamma T}$, γ can be determined by plotting a graph of $\ln(a_n)$ against n with $ln(a_n) = \ln(a_0) - \gamma Tn$. In the case of a) there was approximately no damping, thus γ was zero and the plot consisted in a straight line with a very, very small negative slope, whereas b) showed a straight line with a significantly negative slope.

iv.   Determination of the quality factor Q

Out of the slope of these plots the quality factor Q could be determined as $-\gamma T = -\gamma \left(\frac{2\pi}{\omega_1} \right) = \frac{-\pi}{Q}$ under the assumption that for low damping $\omega_0 \sim \omega_1$. The errors were estimated via linear regression analysis.

5

a) $Q_1 = \frac{-\pi}{m_{exp1}} = 264.00 \pm 20.91$ (no damping)

b) $Q_2 = \frac{-\pi}{m_{exp2}} = 6.02 \pm 1.33$ (with a brake current of 0.6A)

v.   Investigating the effect of increased damping

In order to investigate the effect of increased damping, the current through the eddy brake was increased to the maximum of 2A. The pendulum then moved from rest directly to the equilibrium position, overshot it by 0.2 on the annulus scale and went straight back to its equilibrium position. Hence, the system was almost critically damped and did not oscillate. In general, the more damping, the larger $\gamma = \frac{b}{2I}$ and therefore the less time it takes to decay, as can be mathematically concluded by looking at the equation of motion:

$$\theta = Re\{Be^{pt}\} = \theta_0 e^{-\gamma t} \cos(\omega_1 t)$$

2.) Forced oscillations

i.   Description of initial behavior when switched on

It took around 30s for the pendulum to reach a steady state. When the driving couple was switched on, the pendulum's response appeared rather irregular, as it consists in the sum of the transient response (oscillating at ω_1 and exponentially decaying with time) and the steady state solution (oscillation at ω_d with constant amplitude) – the combination of two oscillations at different frequencies. Only once the transient had died away, it settled down to oscillate with constant amplitude at a frequency ω_d.

ii.   Amplitude of forced oscillations as a function of angular frequency (see VII)

This relationship was investigated at two different damping conditions. In a) the eddy brake current was set to $I_b = 0.3A$ and b) $I_b=0.6A$. Due to the strict time frame of the experiment, the period of the oscillation was measured using a stopwatch whilst waiting for the pendulum to settle down into SHM. The measurements were made over the whole accessible frequency range to find exceptional features of the response and the resonance frequency. Zero offsets were avoided by measuring the amplitude from both sides of zero and taking the average.

iii.   Deductions from plot

As theoretically predicted, both graphs have two limiting values $\lim_{\omega \to 0} X(\omega) = \frac{A/_I}{\omega_0^2} = constant$ and $\lim_{\omega \to \infty} X(\omega) = 0$ at their outer ends while each of them shows one peak embedded in the resonance curve. The amplitude resonant frequency can be calculated to be $\omega_{max} = \sqrt{\omega_0^2 - 2\gamma}$. In the case of very low damping ω_0 is the limiting value for w_max. In general, we

6

can derive from the equation for the steady-state amplitude of a forced oscillation that the higher γ the lower the amplitude, which we verified experimentally with our measurement results. When the braking current was 0.6A instead of 0.3A, the amplitude of the graph was much lower. (See VII)

a) Low damping at $I_b = 0.3A$: $\omega_{max} = 3.30 \pm 0.02 \frac{rad}{s}$; $A_{max\_1} = 16.3 \pm 0.8 \ scale\ units$

b) High damping at $I_b = 0.6A$: $\omega_{max} = 3.30 \pm 0.1 \frac{rad}{s}$; $A_{max\_2} = 4.9 \pm 0.3 \ scale\ units$

As expected, the resonant frequency is very close (within 1σ) to that obtained for the free oscillation:

$$\omega_{res} = 3.30 \pm 0.02 \frac{rad}{s} \sim \omega_1 = \frac{2\pi}{T_1} = 3.28 \pm 0.014 \frac{rad}{s}.$$

The amplitude at $\omega = 0$ was obtained by rotating the drive wheel slowly by hand and reading off the maximum displacement to be 0.65 scale units.

$$Q_1 = \frac{A_{max\_1}}{A_r} = \frac{16.3 \pm 0.8}{0.65 \pm 0.1} = 25.1 \pm 2$$

$$Q_2 = \frac{A_{max\_2}}{A_r} = \frac{4.9 \pm 0.3}{0.65 \pm 0.1} = 7.5 \pm 0.7$$

The second value of Q under a brake current of 6A has already been determined to be 6.02 ± 1.33 by plotting the natural logarithm of the amplitude against the number of oscillations, so that we can verify that both methods lead to the very similar results, in the ranges of their errors even the same result.

If there was no damping the amplitude would be

$$\frac{A/_I}{2\gamma\omega_0} = QX(\omega \to 0) = (264 \pm 21)(0.65 \pm 0.1)scale\ units = 171.6 \pm 20\ scale\ units.$$

In the given physical system, this value overshoots the scale numerous times and would thus lead to the destruction of the torsion pendulum.

IV. Discussion

First of all, we have to consider to what extent it is justifiable not to include damping due to internal friction and the air resistance in our calculations. As the torsion pendulum lasted more than 15 minutes rotating before it stopped at its equilibrium position, this error can be neglected for the purposes of this experiment.

Secondly, we have to ask ourselves if human beings can use a stopwatch well enough to obtain accurate results: The Response Time for college-age individuals is approximately 190 milliseconds for detecting a visual stimulus (the pointer for displacement at the maximal amplitude) and responding to it (pressing a button on the electronic stopwatch).(see VI.4) The stopwatch is likely to be controlled by a quartz crystal with an error of at most a few seconds a day (a few parts in 10^5 or better), so this is unlikely to affect the result, whereas the human observation is likely to induce an error in the results.

Thirdly, the scale of the torsion pendulum only showed integer values, so that most of the observed decimal places in the observation of the angular displacement are estimated within an error range of $\pm 0.2\ scale\ units$. But even if that provides lots of potential errors, there are no zero offsets due to this experimental method as I indicated in section III.

In general, we can deduce from our experimental data that our measurements were very precise and accurate within the range of our possibilities because we obtained two very similar results for the quality factor of a system that is damped by a current of 0.6A through the eddy current brake: Q ~ [4.69, 8.2] by two different methods.

$$Q_{0.6A} = \frac{-\pi}{m_{exp2}} = 6.02 \pm 1.33$$

$$Q_{0.6A} = \frac{A_2}{A_r} = \frac{4.9 \pm 0.3}{0.65 \pm 0.1} = 7.5 \pm 0.7$$

The first approach was based on measuring the maximum displacement for each successive oscillation and deducing the slope from the plot of the natural logarithm of the amplitude against the number of oscillations (see III.1.ii-iv), whereas the second estimate of Q was obtained under forced oscillation conditions by taking the ratio of the experimentally determined resonance amplitude and the amplitude of natural oscillation (see III.2.iii).

While Q measures the persistence of the oscillation in terms of the unforced motion (free decay, transient response), it measures the height and sharpness of the resonance in terms of the forced motion (steady state response).

V.   Conclusions

With this experiment, the theoretical expectations were clearly verified:

1.) As $Q = \frac{\omega_0}{2\gamma}$, it is antiproportional to the damping. Thus the higher is the damping the lower is Q as we could deduce from our data. (see III.1.iii and III.2.iii)
2.) We were also able to verify that the critically damped as well as the overdamped case do not oscillate. (see III.1.v)

3.) Moreover, observing the initial behavior of the simultaneously damped and forced system, we were able to justify that the general solution of equation (2) consists in the sum of the transient response and a particular solution – the steady-state response. (see III.2.i)

4.) The resonant frequency was very close within 1σ to that obtained for the free oscillation:

$$\omega_{res} = 3.30 \pm 0.02\,\frac{rad}{s} \sim \omega_1 = \frac{2\pi}{T} = 3.280 \pm 0.014\,\frac{rad}{s}$$

as predicted in theory. (see III.2.iii)

5.) Both graphs have two limiting values $\lim_T(\omega\to 0)[X(\omega)]=(A/I)/(\omega\_0^2)=$constant and $\lim_T(\omega\to\infty)\,[X(\omega)]=0$ at their outer ends and their resonance frequencies are very close to each other as expected for a small difference in the damping constant γ.

VI.   References

1. NST Part 1A Physics Practicals Class Manual, Lent Term 2011, pp7-15
2. http://ftp.first-tale.de/Uni/SS08/Labor%201/Pohlsches%20Drehpendel/Lab1V08-PohlschesRad.pdf
3. http://www.npl.washington.edu/eotwash/experiments/experiments.html
4. http://biae.clemson.edu/bpc/bp/Lab/110/reaction.htm#Type%20of%20Stimulus

VII.   Appendix

a) Plot of amplitude of forced oscillations as a function of angular frequency